The Magnetic Field

Johannes G. Lang

SIEMENS AKTIENGESELLSCHAFT
HEYDEN & SON

Heyden & Son Ltd., Spectrum House, Hillview Gardens, London NW4 2JQ, UK.
Heyden & Son Inc., 247 South 41st Street, Philadelphia, Pennsylvania 19104, USA.
Heyden & Son GmbH, Münsterstrasse 22, 4440 Rheine, West Germany.

Title of German original edition:
Das Magnetische Feld (PU 07)
Von Johannes G. Lang
Siemens Aktiengesellschaft 1974
ISBN 3-8009-4007-8

ISBN 3-8009-4707-2 Siemens AG, Berlin and München
ISBN 0-85501-503-9 Heyden & Son Ltd, London

Typeset in Great Britain by The Universities Press, Belfast
Printed in Great Britain by Henry Ling Limited, The Dorset Press, Dorchester

Introduction

This programmed instruction (PI) book describes in a straightforward and logical manner the *magnetic field.* A magnetic field is always present when an electric current is flowing and is important in electrical apparatus and machines. This book is intended principally for use at technical college level. It is also suitable for secondary school use, as a refresher for students of advanced technology and for adult education.

When you have successfully worked through the book you will be familiar with:

The meaning of ferromagnetism, magnetic flux, magnetic flux density, magnetic saturation and residual magnetism;
SI units of magnetic flux and magnetic flux density;
The rules relating to the direction of lines of magnetic flux and to the polarities of current-carrying solenoids;
The interaction between magnetic fields and the electrodynamic effect.

The programmed instruction method means that the material is divided into small steps and by answering the questions asked after each step you can easily check your progress. You can work at your own speed and if you do not understand a step you should repeat it until it is clear.

You should use the book in the following manner:

1. Work thoroughly through each Lesson.
2. Answer the Question at the end of each lesson.
3. Check that your Answer agrees with that given on the next page. If it does not agree work through the lesson again and find out where you made a mistake.
4. Answer the questions in the Intermediate Tests and the Final Test.
5. Check your answers with those provided and repeat any lessons that you did not fully understand.

Some of the technical terms used in this book are explained briefly in the Appendix (p. 66).

London, February 1978

THE PUBLISHERS

Properties of magnets

Iron (normally in the form of mild steel or cast iron), nickel, cobalt, certain alloys, e.g. Alnico (aluminium–nickel–cobalt) or sintered oxide mixtures (Ferrites) are referred to as ferromagnetic materials; they are attracted by magnets.

By treating a piece of ferromagnetic material in a special manner, it is possible to convert it into a permanent magnet.

The most commonly known forms of magnets are the bar, the horse-shoe and the compass needle.

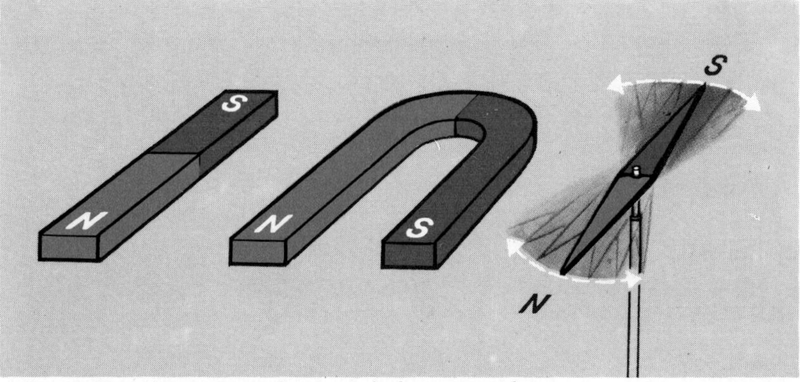

All magnets have two characteristic properties:

1. They attract and hold iron (force effect)
2. If free to move they assume a roughly north–south position (directional effect).

The directional effect is utilized in the compass. The pivoted compass needle takes up a position in the magnetic field of the earth so that the same point (usually coloured blue) always points roughly towards the geographical north pole. This point of the compass needle is called the north pole; the opposite point the . . . .

QUESTION 1

Add the missing words.

If you know the answer write it down in your exercise book (not in the instruction book!). First write Answer, then the number of the lesson as printed in the top right-hand corner and then your answer, thus:

Answer 1 . . . .

Compare your answer with the solution on the other side of this page only after having answered the question.

The answer in your exercise book should read:

Answer 1

South pole

No doubt your answer was correct and so your exercise book now contains the first correct entry. Enter subsequent answers in it, in a similar way, one beneath the other.

A bar magnet dipped into iron filings will attract some of them. The upper illustration shows what the pattern made by the filings will look like.

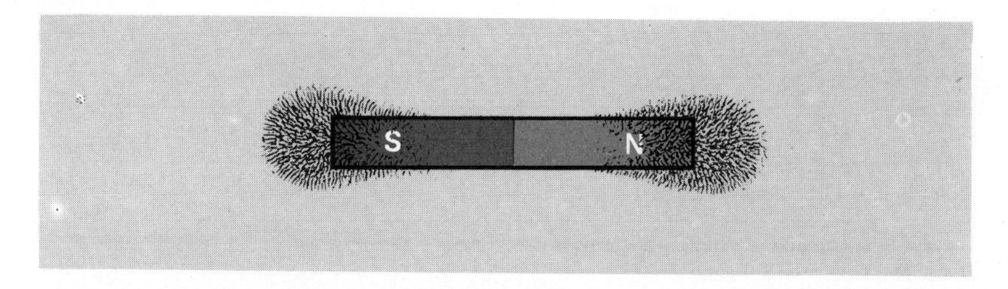

Most iron filings will adhere to the two ends of the bar magnet while hardly any are attracted to the centre part. The ends of the bar magnet, which attract most of the iron filings, are called the *poles* of the magnet. Every magnet has one north pole and one south pole, distinguished by their directional effects.

If we examine the interaction between *two* magnets, we find that forces are acting between their poles, as shown below:

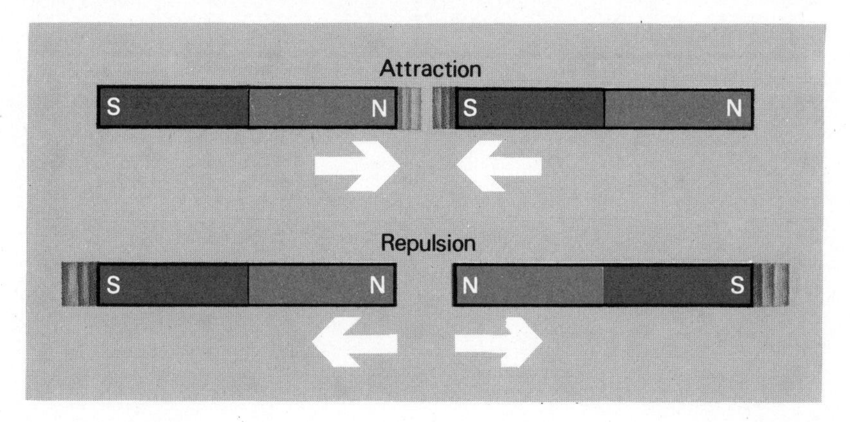

QUESTION 2

From an examination of the illustration state the law which relates to the force of attraction or repulsion acting between the poles of two magnets and note it down in your exercise book.

Answer 2

Poles of opposite polarity (i.e. north and south poles) attract each other.

Poles of the same polarity (i.e. two north or two south poles) repel each other.

If we cut a bar magnet in half, new poles of opposite polarity will form at the two surfaces of the cut, where there were none before. Each half of the bar magnet now has a north pole and a south pole.

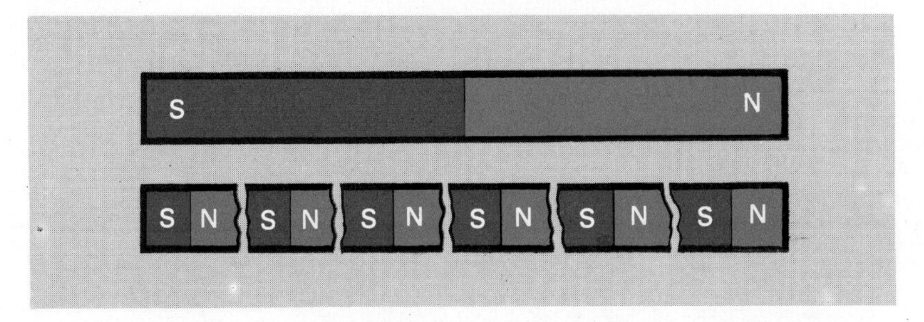

We can continue to split the bar magnet into smaller and smaller pieces yet even the most minute parts into which we split it (i.e. its molecules) will still be magnetic. These smallest magnetic particles are called *molecular magnets.*

All magnetic materials are formed from such molecular magnets. So long as the material is not magnetized, these molecular magnets are *randomly oriented* and produce no external magnetic effect. This can be changed by bringing a permanent magnet into contact with the ferromagnetic material which thereby becomes magnetized due to more and more of its molecular magnets aligning or orienting themselves so that their north and south poles face in the same direction; the material becomes magnetic.

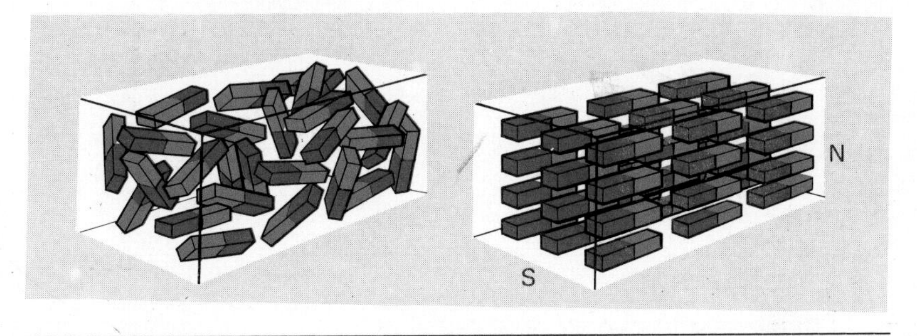

QUESTION 3

What happens *inside* the material when it is magnetized?

Answer 3

Magnetization causes all molecular magnets to align themselves and to form a common north pole and a common south pole.

Magnetic saturation and residual magnetism **Lesson 4**

The more the molecular magnets are aligned inside a magnetic material, the stronger its magnetic effect. Once all its molecular magnets are aligned, no further increase of its magnetic effect is possible. The material has reached *magnetic saturation.*

Some materials retain their magnetism fairly well after being magnetized. Only a small number of molecular magnets revert to their disoriented state once the magnetizing effect has been removed. The majority of the molecular magnets remain in their aligned positions, held there by internal forces. Such materials are called *hard magnetic materials.*

Materials, which lose most of their magnetism once the magnetizing effect has been removed, are called *soft magnetic materials.* Their internal forces are small, the molecular magnets are easily aligned, and will readily resume their disoriented state. Dynamo sheet steel is an example of a soft magnetic material. This characteristic, amongst others, makes sheets of soft magnetic materials suitable for transformer cores.

The magnetism which remains in a ferromagnetic material after the source of magnetization has been removed is called the *residual magnetism* or *remanence.* Severe shocks or heating can partially or wholly destroy the alignment of the molecular magnets of both hard and soft magnetic materials.

QUESTION 4

What can cause the weakening of the magnetic properties of a permanent magnet?

Answer 4

The magnetism of a permanent magnet is partially or wholly removed by severe shocks or heating.

The space in which magnetic forces act is called a *magnetic field*. Such a field exists, for example, between the ends of a bar magnet or the arms of a horse-shoe magnet.

The effects of magnetic fields, like electric fields† can be made visible. A sheet of paper stretched across a frame is placed over the magnet and iron filings are loosely scattered over it; they will arrange themselves along lines under the influence of the magnetic field. Therefore, we talk about magnetic field lines and we must imagine that the space around a magnet is filled with such lines.

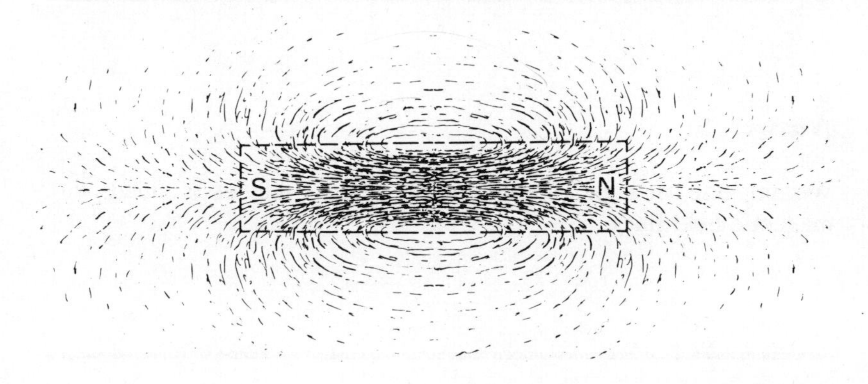

The illustration shows the magnetic field pattern in which iron filings will arrange themselves around a bar magnet.

† See instruction programme PI02 'The Electric Field'.

QUESTION 5

How do we imagine the space around a bar magnet?

Answer 5

We imagine that the space around a bar magnet is filled with magnetic field lines.

If we draw the way the iron filings have arranged themselves around the bar magnet in Lesson 5 we obtain the following picture:

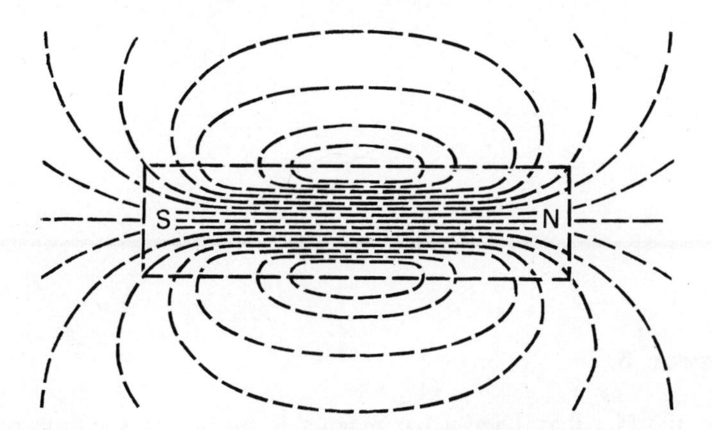

The broken lines show the paths of the magnetic field lines. It is sufficient to show just a few of these field lines in order to picture the magnetic field.

The field lines do not only exist outside the magnet, they also continue inside it. This leads to the following rule:

The lines of a magnetic field always form closed loops

The lines of a magnetic field are called the lines of *magnetic flux.*

QUESTION 6

From what fact, stated in an earlier lesson, can we deduce that there must be lines of magnetic flux inside the material of a magnet?

Answer 6

From the fact that when a bar magnet is cut up the separate parts themselves become magnets (see Lesson 3); every magnet thus consists of molecular magnets.

Intermediate Test 1

The questions in this test deal with the subject matter of the preceding lessons. If you are unable to answer all the questions you are advised to work through the lessons again.

1 How does the force effect of a magnet manifest itself?

2 How does one explain saturation of ferromagnetic materials?

3 Why are soft magnetic materials unsuitable for permanent magnets?

Answers to Intermediate Test 1

1 Iron or other ferromagnetic materials are attracted and hold fast to the magnet.

2 A magnetic material is saturated when all its molecular magnets are correctly aligned.

3 The molecular magnet of a soft magnetic material can move easily; it can therefore also easily lose its magnetism.

Magnetic fields exert forces (see Lesson 2). These forces are the stronger the more lines of magnetic flux there are in the magnetic field or the closer these lines are bunched together.

The maximum force between two bar magnets acts directly between their poles (attraction or repulsion) because this is where the density of the magnetic flux is at its greatest.

The *magnetic flux density* is a measure of the effect exerted by a magnetic field. It is also called the magnetic induction.

The *magnetic flux density* specifies the value of the magnetic flux passing normally through, i.e. at right-angles to, a unit area (cm^2 or m^2) of the magnetic field.

QUESTION 7

How does the external magnetic flux density of a bar magnet vary with increasing distance from the poles?

Answer 7

The magnetic flux density decreases with increasing distance from the poles.

Electric current and magnetic field — Lesson 8

Magnetic fields are not necessarily only associated with magnetic materials. Every electric current generates a magnetic field without any magnetic material being present. A simple experiment will prove this:

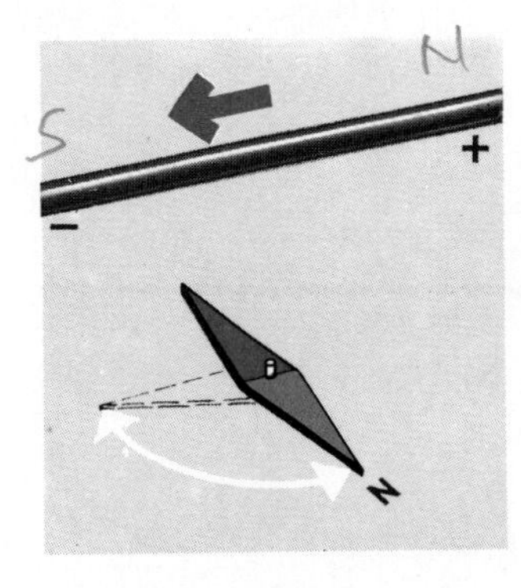

Let a direct current flow through a horizontal, straight piece of wire, preferably oriented in a north–south direction. A compass needle placed near the wire will be deflected from its natural rest position as long as the current flows. The force acting on the compass needle is present along the whole length of the wire. This means:

A current-carrying conductor generates a magnetic field in its vicinity.

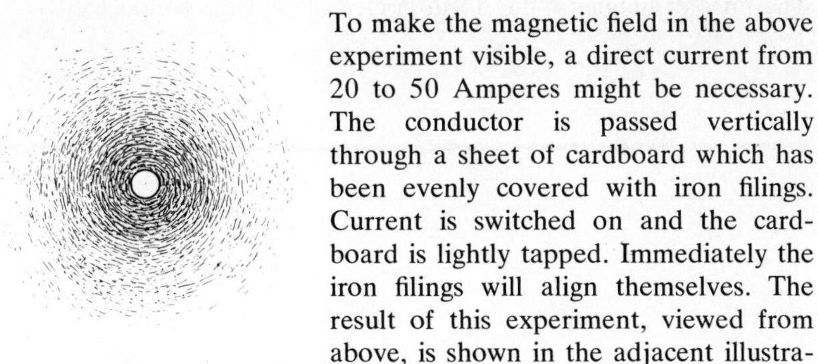

To make the magnetic field in the above experiment visible, a direct current from 20 to 50 Amperes might be necessary. The conductor is passed vertically through a sheet of cardboard which has been evenly covered with iron filings. Current is switched on and the cardboard is lightly tapped. Immediately the iron filings will align themselves. The result of this experiment, viewed from above, is shown in the adjacent illustration.

QUESTION 8

What pattern do the lines of magnetic flux in this experiment assume?

Answer 8

The lines of magnetic flux form circles around the conductor.

Field distribution around a current-carrying conductor — Lesson 9

The lines of magnetic flux around a current-carrying conductor form closed loops, i.e. circles; their common centre is the centre of the conductor.

The magnetic field extends through the space along the whole length of the conductor; lines of magnetic flux close together may give the appearance of tubes of flux around the conductor. Magnetic fields appear around every current-carrying conductor, even around liquid or gaseous ones.

The magnetic flux density is a maximum at the surface of the conductor and decreases with the distance from it. It is quite immaterial whether the conductor is bare or insulated since the field in any non-magnetic material is roughly the same as in air.

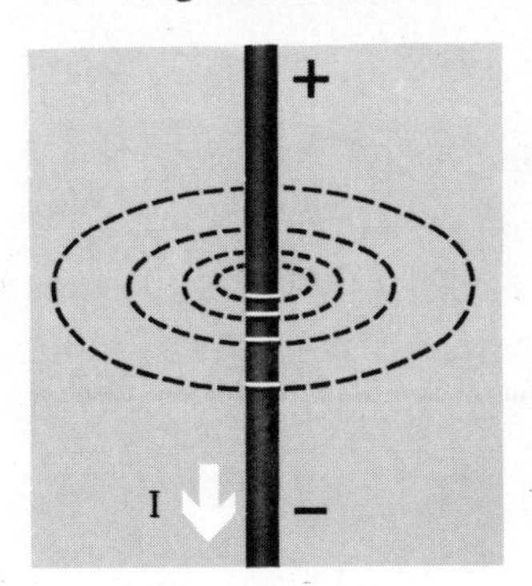

The illustration shows the lines of magnetic flux in only one plane which is perpendicular to the current-carrying conductor. Circles are drawn at such distances that the same number of lines of magnetic flux lie between any one circle and the next circle. The distance between two circles increases with the distance from the conductor, indicating the decreasing magnetic flux density.

QUESTION 9

What property of the magnetic field can we deduce from the appearance of the lines of magnetic flux around a magnet or current-carrying conductor?

Answer 9

The appearance of the lines of magnetic flux shows that they always form closed loops.

Direction of the magnetic field **Lesson 10**

A compass needle moved in a circular path around a conductor carrying a sufficiently heavy current, will always assume a position at right angles to the radius, thus indicating the direction of the lines of magnetic flux.

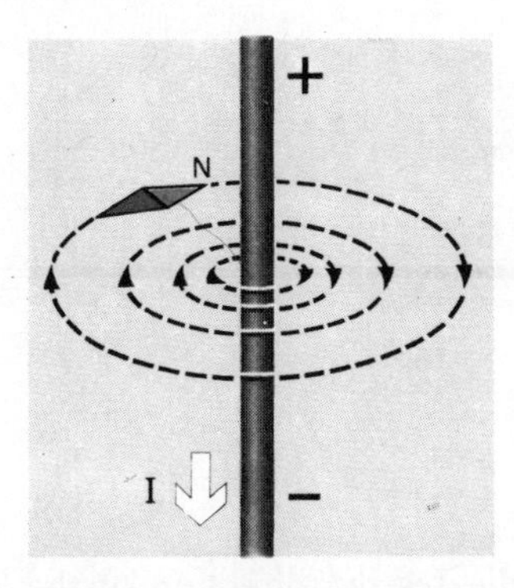

By convention: the *north pole* of a compass needle always indicates *the direction* of the lines of magnetic flux, i.e. the magnetic field.

QUESTION 10

How can one use a compass needle to determine the direction of a magnetic field?

Answer 10

The north pole of the compass needle indicates the direction of the magnetic field.

Based on the above convention, an observation of the position of the compass needle (from Lesson 8) indicates that a connection must exist between the direction of current flow and the direction of the magnetic field.

However, before we examine this connection in detail let us first consider the following:

Electrons moving in a conductor constitute a current. Even before the discovery of electrons (about 1895) a convention had been agreed upon concerning the direction of a current. According to this, an electric current flows from the positive pole (+ terminal) of a source of electricity through the conductor to the load, and then through the return conductor back to the negative pole (− terminal). (See left-hand illustration).

The conventional direction of current flow is opposite to the direction of electron movement.

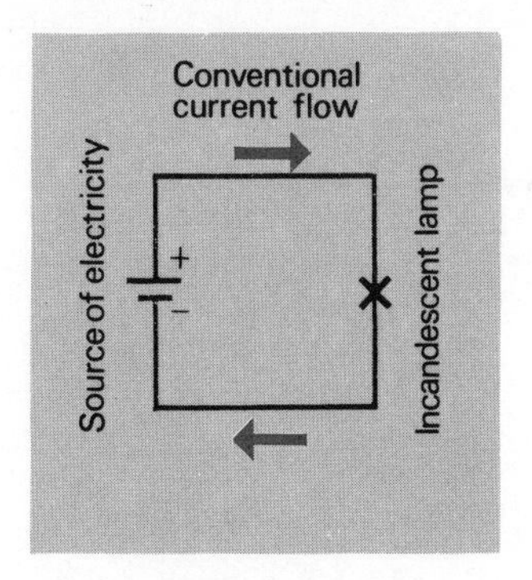

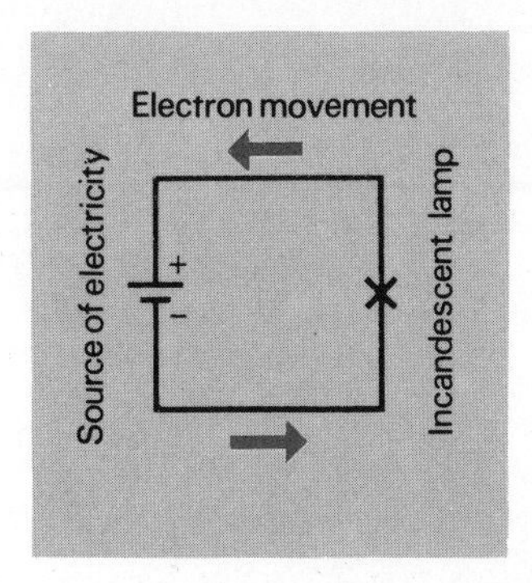

The conventional direction of current flow is used throughout the following pages.

QUESTION 11

What position will a freely movable compass needle take up when placed between the poles of a horse-shoe magnet?

Answer 11

The compass needle assumes a position with its north pole pointing towards the south pole of the horse-shoe magnet. (Opposite poles attract).

Indicating the direction of current flow Lesson 12

A cross and a dot are used to indicate pictorially the direction of current flow in a conductor. These symbols are derived from the picture of an arrow in flight.

Looking in the direction of the arrow as it flies away one sees the ends of its flights; these can be depicted schematically by a *cross*.

If the arrow is coming towards you, you would see the point of the arrow; this is depicted by a *dot*.

This convention is used in the illustration below to indicate the directions of current flow in two current-carrying conductors.

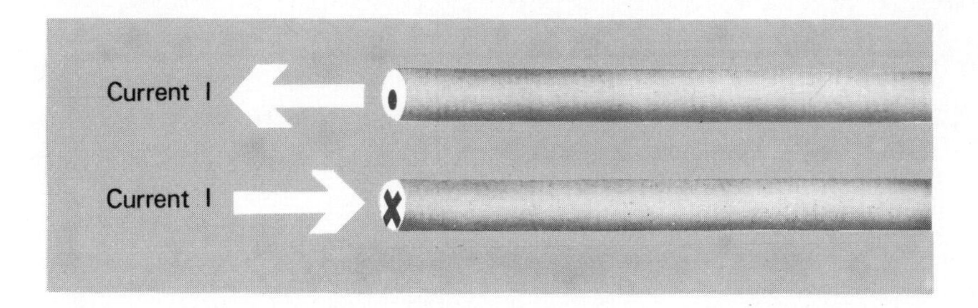

QUESTION 12

Which of these three statements is correct?

The lines of magnetic flux in the vicinity of a bar magnet

(a) run from the south pole to the north pole

(b) run from the north pole to the south pole

(c) run in either direction from pole to pole

Answer 12

(b) The lines of magnetic flux run from the north pole to the south pole.

Intermediate Test 2

1 Why is the magnetic flux density (magnetic induction) an important characteristic of a magnetic field?

2 Do magnetic fields form around liquid or gaseous conductors?

3 How does the magnetic flux density around a current-carrying conductor change with increasing distance from the conductor?

(a) it remains constant

(b) it decreases

(c) it increases

Answers to Intermediate Test 2

1 Because the magnetic flux density is a measure of the strength of a magnetic field.

2 Every current-carrying conductor generates a magnetic field.

3 (b) it decreases

Direction of current flow and direction of field

Lesson 13

The illustration below shows the connection between the direction of current flow and the direction of its magnetic field.

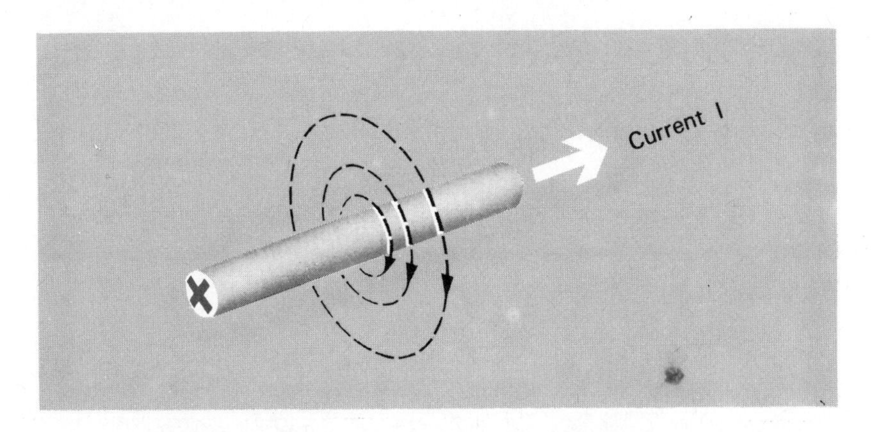

An arrow is used in place of a compass needle; its point indicates the direction of the lines of magnetic flux.

> *Looking in the direction of current flow the lines of magnetic flux around the conductor are in a clockwise direction.*

QUESTION 13

Determine the direction of the lines of magnetic flux in the illustration below and draw the completed illustration in your exercise book.

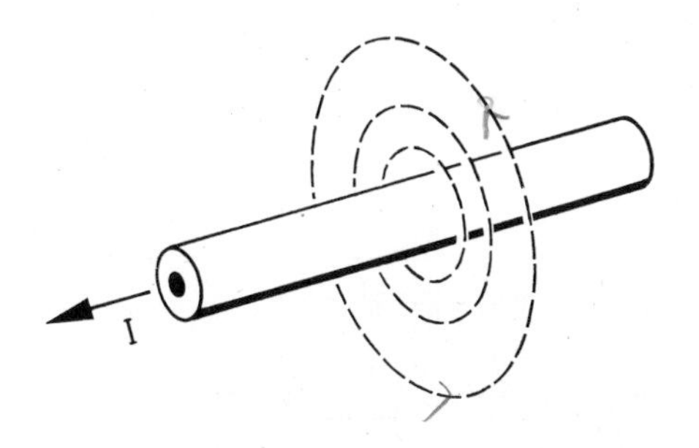

Answer 13

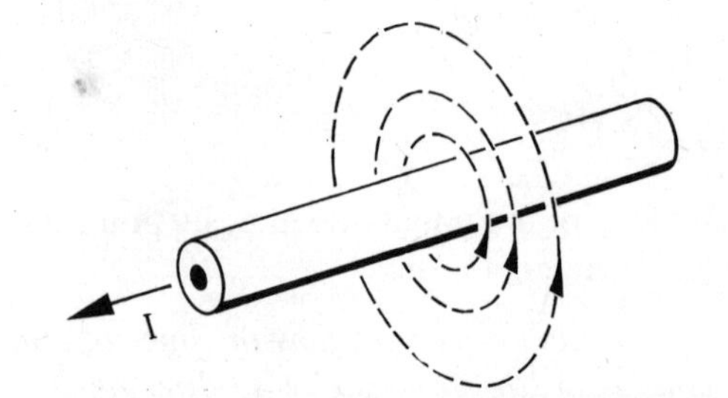

The connection between the direction of current flow and the direction of its lines of magnetic flux can be memorized by a number of rules. One of these is the 'cork-screw rule' (right-handed screw rule).

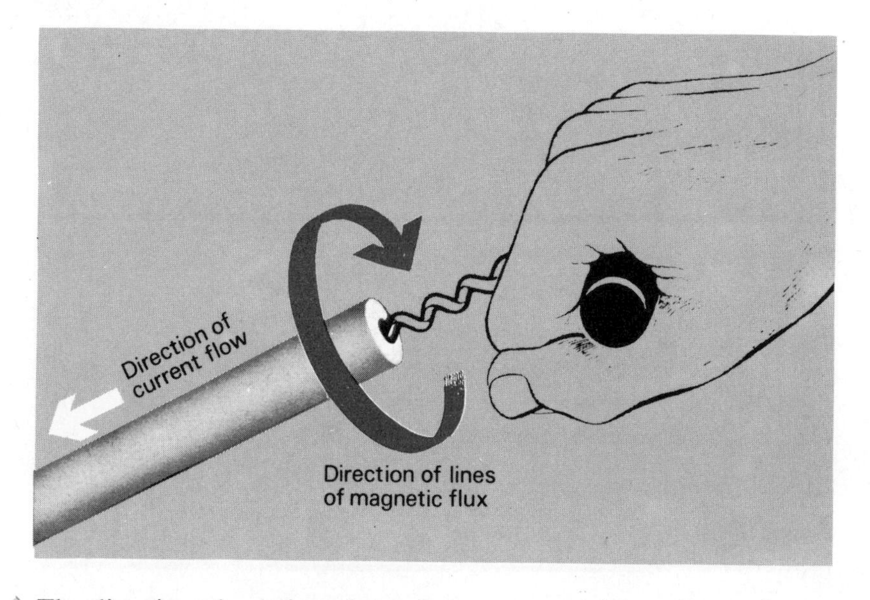

The direction of rotation of a cork-screw screwed *into* the conductor in the direction of current flow indicates the direction of the lines of magnetic flux around the conductor.

QUESTION 14

Do you consider that a complete picture is obtained (in the illustration relating to Lesson 13) if the lines of magnetic flux are shown only at one point along the conductor. Explain your reasoning.

Answer 14

Not really, because if one imagines the lines of magnetic flux along the conductor they would form tubes of magnetic flux around the conductor and concentric with it.

Force between a magnetic field and a current-carrying conductor

Lesson 15

The experiment described in Lesson 2 proves that forces act between the magnetic fields of two magnets. Forces also act if one of these magnetic fields is generated around a current-carrying conductor. A simple experiment will prove this:

A conductor suspended from thin flexible metal tapes is placed in the field of a horseshoe magnet.

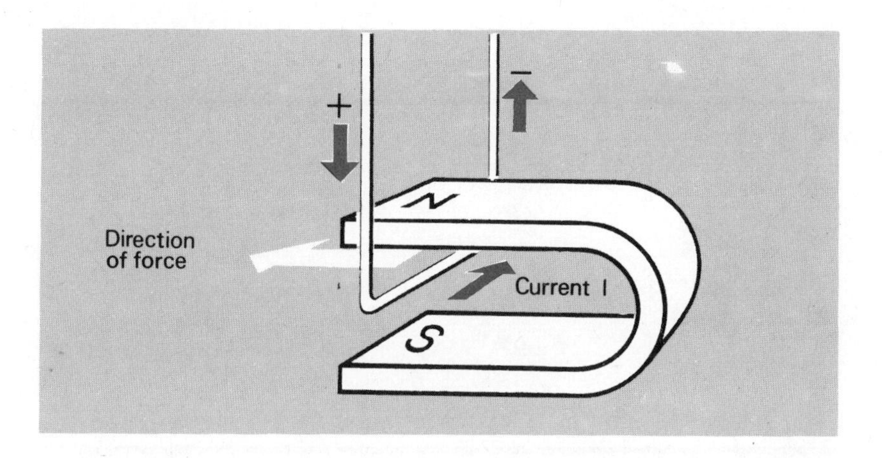

The two leads marked + and − are connected to the poles of a source of electricity. As soon as current flows the magnetic fields produced by the magnet and by the current interact and the conductor is deflected.

QUESTION 15

What kind of effect does a magnetic field exert on a current-carrying conductor?

Answer 15

A force effect.

To understand how this force is generated, one has to consider the magnetic field resulting from the reaction between the magnetic field of the current-carrying conductor and the magnetic field of the magnet. In the illustration below, the magnetic field of the magnet is shown on the left-hand side; that around the conductor in the centre and the combined magnetic field on the right-hand side.

Any opposing lines of magnetic flux from the magnet and from the conductor will wholly or partially cancel each other. The combined magnetic field is thus weakened to the left of the conductor (right-hand illustration). To the right of the conductor all the lines of magnetic flux point in the same direction, thus increasing the strength of the magnetic field on that side.

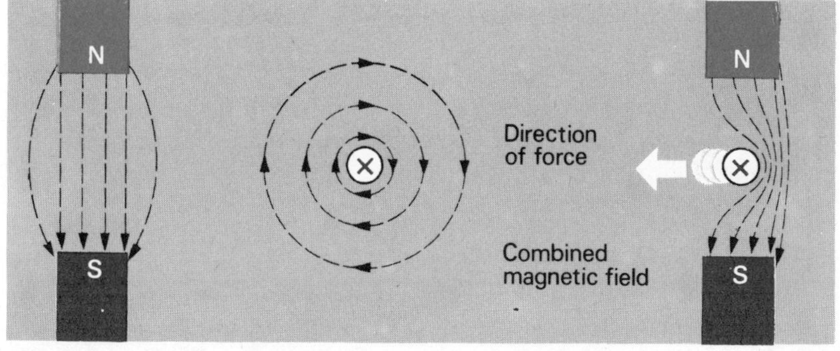

There is a tendency for this unequal magnetic field distribution on the two sides of the conductor to revert to a uniform one. Lines of magnetic flux have a tendency to shorten themselves. This results in the movable conductor being pushed from the stronger to the weaker region of the magnetic field. This phenomenon is called the *electrodynamic* effect.

Remember: *A current carrying conductor is forced from the stronger to the weaker part of the magnetic field.*

The electrodynamic force increases in proportion to the magnetic flux density of the magnetic field, to the current flowing through the conductor and to the effective conductor length.

QUESTION 16

What changes are required in the experiment in Lesson 15 to cause the current-carrying conductor to be deflected in the opposite direction?

Answer 16

To make the current-carrying conductor move in the opposite direction it is necessary to reverse either the polarity of the magnet or the direction of current flow in the conductor.

The 'left-hand rule' is used to determine the direction of the resulting force. It says:

> *The left-hand is held flat so that the lines of magnetic flux are perpendicular to the palm and directed towards it and the fingers point in the direction of the current flow; the outstretched thumb will then indicate the direction of the deflection.*

The direction of the deflection is at right angles to the lines of magnetic flux and to the direction of current flow.

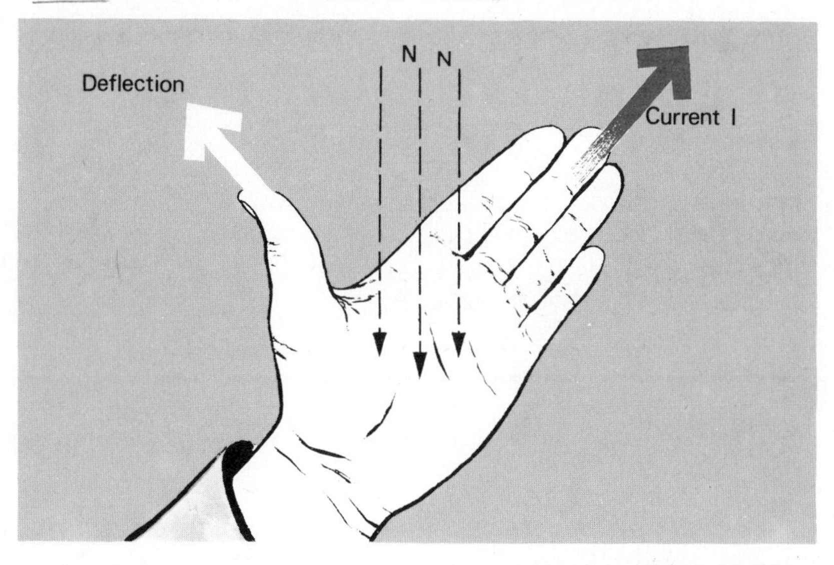

QUESTION 17

Use the 'left-hand rule' to determine the direction in which the conductor will be deflected in the arrangement illustrated below.

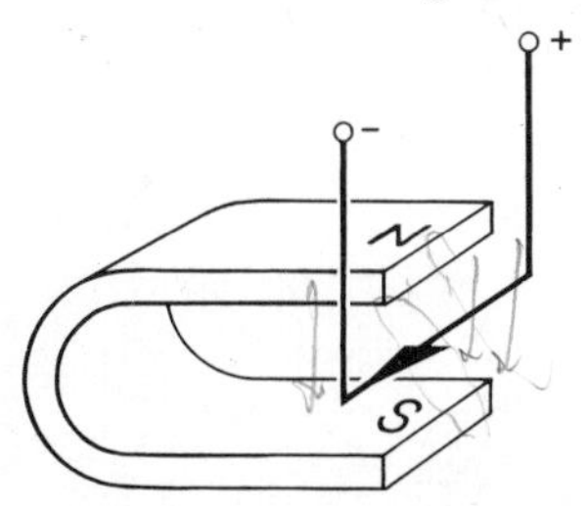

Answer 17

According to the ‘left-hand rule’ the electrodynamic effect will cause the conductor to be expelled from the magnetic field, i.e. deflected to the right.

Forces acting between current-carrying conductors — Lesson 18

An electrodynamic force will also act between two current-carrying conductors.

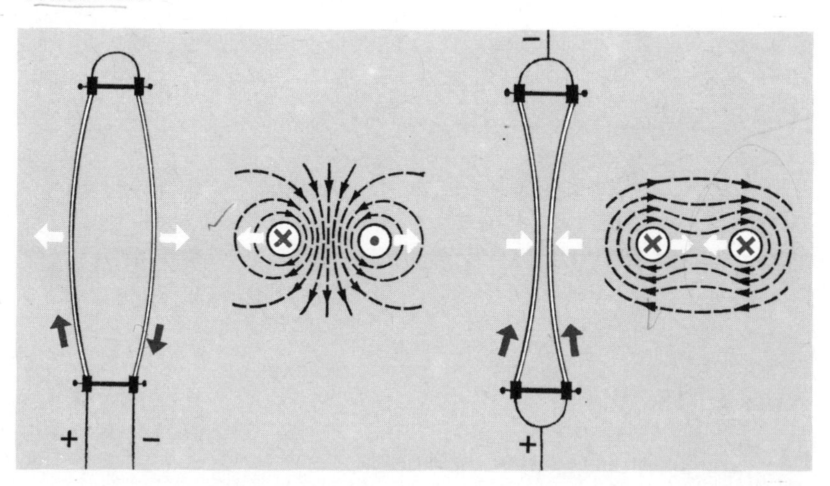

The illustration shows the magnetic field distribution in the region between two current-carrying conductors.

Currents *flowing in opposite directions* generate magnetic fields which are in the same direction between the conductors and where they produce an increase in magnetic field strength. To the left and right of the conductors the two sets of magnetic lines of flux act in opposition to each other, thus causing a weakening of the common magnetic field.

Currents *flowing in the same direction* generate magnetic fields which act in opposite directions between the conductors and where they produce a weakening of the magnetic field strength. To the left and right of the conductors the two sets of magnetic lines of flux act in the same direction, thus causing a strengthening of the common magnetic field.

These uneven flux distributions cause forces to act on the two current-carrying conductors.

QUESTION 18

What are the forces called that act on two current-carrying conductors?

Answer 18

They are called electrodynamic forces.

The electrodynamic forces acting on two current-carrying conductors are the stronger, the larger the currents and the closer the two conductors. This is the reason why bus-bars and windings carrying heavy currents have to be well braced so that they are not distorted by electrodynamic forces.

Try to determine the directions of the electrodynamic forces acting in the illustrations of Lesson 18 and complete the following sentences:

Currents flowing in the same direction in two straight parallel conductors cause the conductors . . . .

Currents flowing in opposite directions in two straight parallel conductors cause the conductors . . . .

QUESTION 19

Fill in the missing words and enter the completed sentences into your exercise book.

Answer 19

. . . to be pulled together.
. . . to be pushed apart.

Intermediate Test 3

1 What is the name of the rule which determines the direction of the force which acts on a current-carrying conductor in a magnetic field?

2 In which direction does the force act on a current-carrying conductor in a magnetic field?

(a) in the direction of current flow

(b) in the opposite direction to current flow

(c) at right angles to the direction of current flow.

3 How can we increase the force acting on a current-carrying conductor in a magnetic field?

Answers to Intermediate Test 3

1 Left-hand rule (Lesson 17)

2 The force acts at right angles to the direction of current flow (Lesson 17)

3 Either by increasing the current flowing through the conductor or by increasing the magnetic flux density of the magnetic field (e.g. by using a stronger magnet). (Lesson 16)

Magnetic field generated by a conductor loop

Lesson 20

In the preceding lessons we showed that an electric current flowing through a conductor generates a magnetic field extending in space over the whole length of the conductor.

For practical applications, however, this magnetic field will not be strong since its magnetic flux density will be too low. It can be strengthened by increasing the current.

The field can also be strengthened by concentrating the field lines. If we form the conductor into a circular loop then all the lines of magnetic flux inside the loop act in the same direction and will increase the magnetic flux density; this is shown in the illustration below:

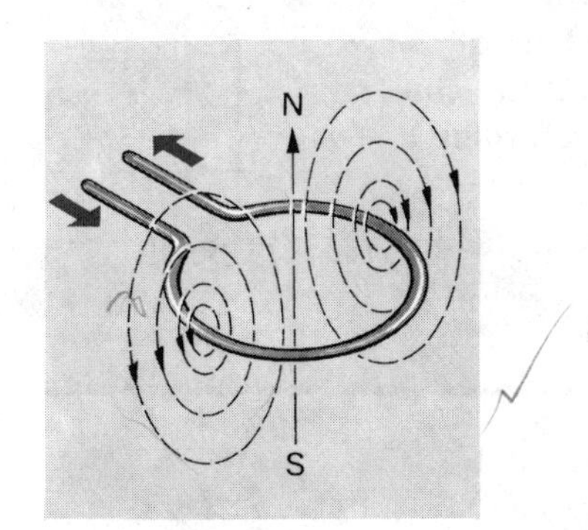

QUESTION 20

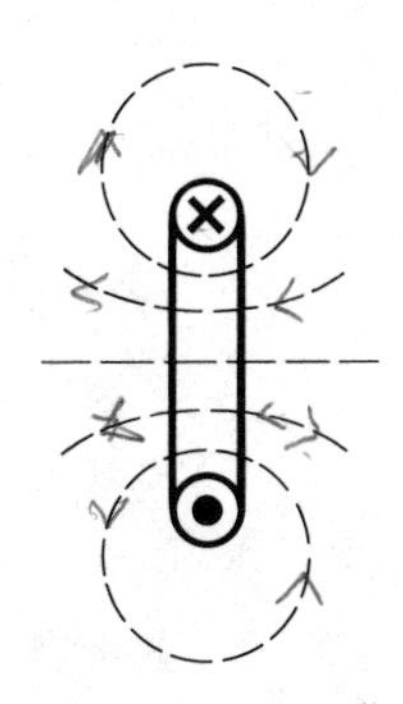

Draw in your exercise book the section through the current-carrying conductor loop shown in the adjacent illustration:

(a) Indicate the direction of the lines of magnetic flux around the conductor (shown by broken lines) by adding arrowheads.

(b) Explain the reason for the increased magnetic field inside the conductor loop.

Answer 20

(a) With the current directions as shown, the lines of magnetic flux leave the plane of the loop towards the left, curve round the conductor and complete the loop by returning to their starting points.

(b) The strengthening of the magnetic field is due to all the lines of magnetic flux inside the loop pointing in the same direction.

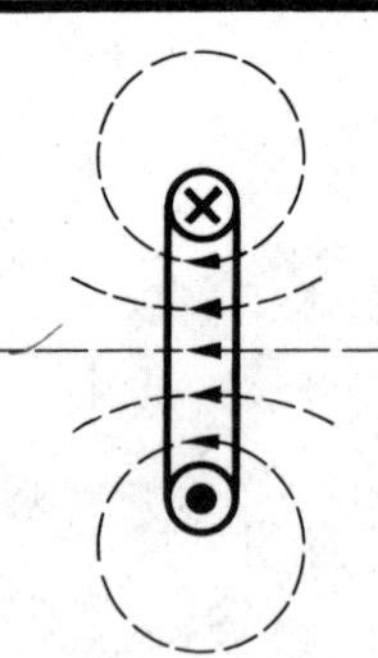

The illustration below shows the field distribution inside a solenoid formed by several turns of wire.

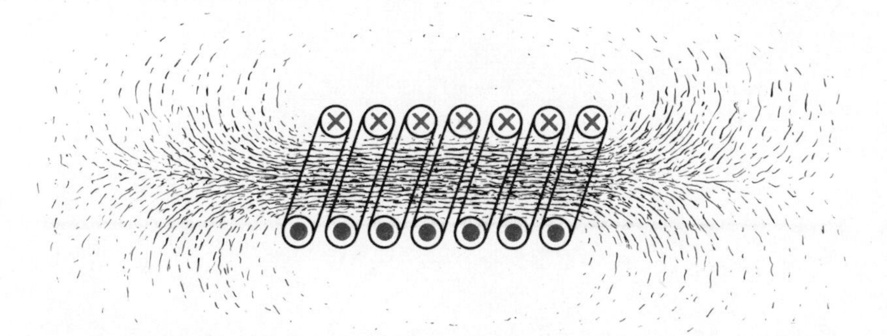

The magnetic fields of the individual turns add up to give the overall magnetic field of the solenoid.

Inside the solenoid the lines of magnetic flux are closely bunched and straight. They are all equidistant. *Outside* the solenoid the lines of magnetic flux open out and then close in a long loop. The magnetic flux density inside the solenoid is thus much greater than on its outside. Furthermore, the internal magnetic field is evenly distributed; it is uniform.

The illustration below is the simplified commonly-used view of the magnetic field of a solenoid. The number of lines of magnetic flux shown is arbitrary.

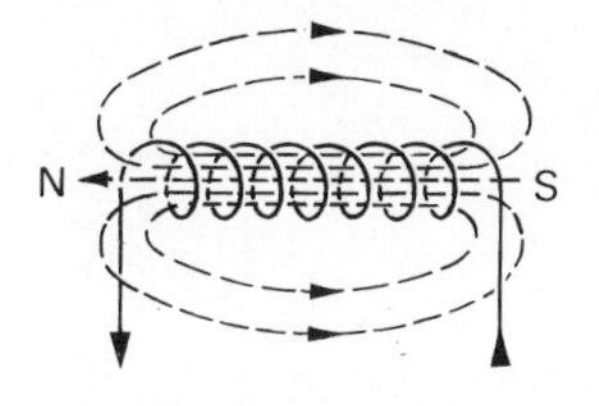

QUESTION 21

What form of magnet has the same magnetic field distribution as a solenoid?

Answer 21

A bar magnet has the same magnetic field distribution as a solenoid (see Lesson 6).

The magnetic field generated by a solenoid is thus similar to that of a bar magnet. The end of the solenoid from which the lines of magnetic flux emanate is called the north pole (as in the case of the bar magnet); the end where they re-enter the south pole.

The ends of a solenoid from which the lines of magnetic flux emerge and re-enter, i.e. its north and south poles respectively, can be readily determined by the *clock rule.*

> *If the current flows through the solenoid in a clock-wise direction we see its south pole.*
>
> *If the current flows in a counter-clockwise direction we see its north pole.*

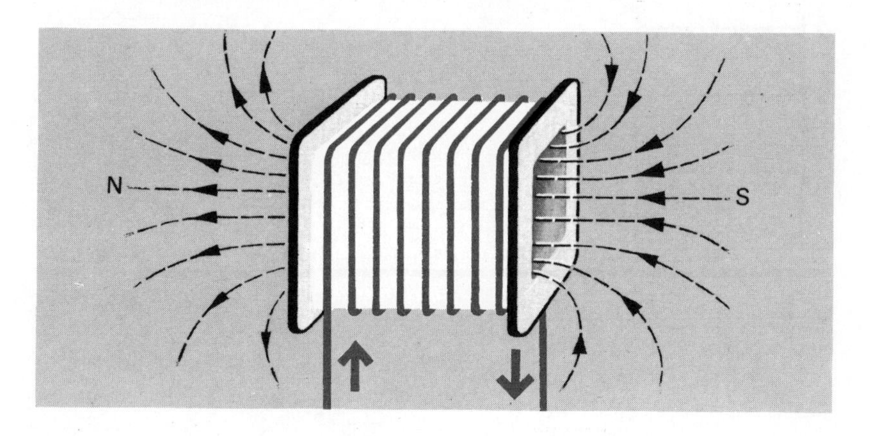

QUESTION 22

Determine the south and north poles of the solenoid shown sectioned in the adjacent illustration; transfer the drawing to your exercise book and complete it.

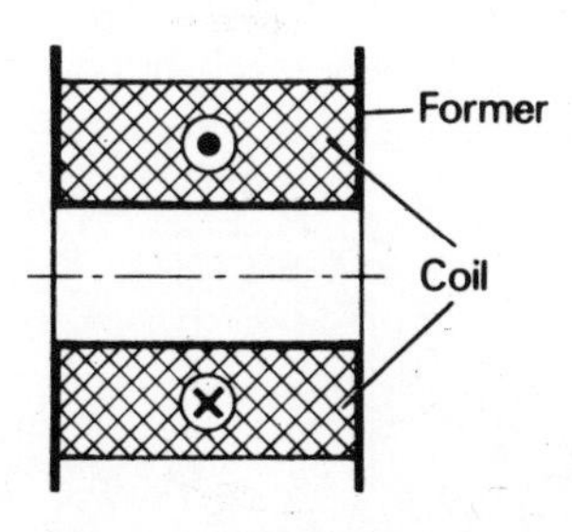

Answer 22

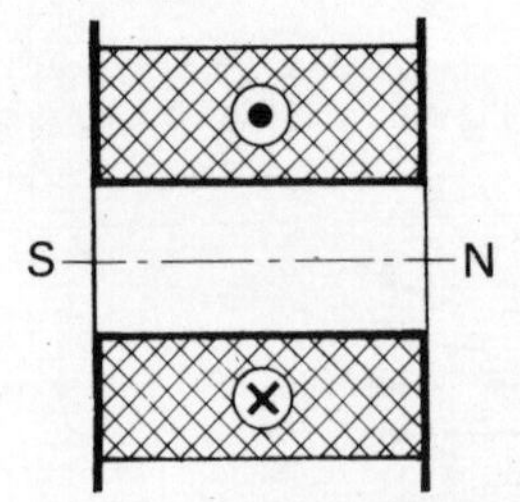

A current-carrying solenoid produces magnetic flux in the same way as a permanent magnet.

The usual symbol for magnetic flux is the Greek capital letter Φ (pronounced phi).

The associated SI-unit is the Weber (Wb). The equivalent unit volt-second (Vs) may be used in place of the Weber.

$$1\ \text{Wb} = 1\ \text{Vs}$$

Another unit for magnetic flux, though no longer used, was the Maxwell (M).

When reading earlier text books the following conversions will be required:

$$1\ \text{Wb} = 10^8\ \text{M}$$

or

$$1\ \text{M} = 10^{-8}\ \text{Wb}$$

QUESTION 23

Convert a magnetic flux Φ of 4×10^4 Maxwells into Webers.

Answer 23

$\Phi = 4 \times 10^4\ \text{M} = 4 \times 10^4 \times 10^{-8}\ \text{Wb} = 4 \times 10^{-4}\ \text{Wb}$

SI unit of magnetic flux density

The magnetic flux density of a uniform magnetic field is readily determined by:

$$\text{Magnetic Flux density} = \frac{\text{Magnetic Flux}}{\text{Area of field}}$$

$$B = \frac{\Phi}{A}$$

The symbol B is used for magnetic flux density.

The SI Unit of magnetic flux density is the Tesla (T)

$$1\ \text{Tesla} = 1\ \text{Weber/m}^2$$

Earlier text books used the unit Gauss (G) for magnetic flux density

$$1\ \text{G} = 1\ \text{M/cm}^2$$

We thus have the following conversion:

$$1\ \text{T} = 1\ \text{Wb/m}^2 = \frac{10^8}{10^4}\ \text{M/cm}^2 = 10^4\ \text{M/cm}^2 = 10^4\ \text{G}$$

QUESTION 24

Example:

Given that: flux $\Phi = 4 \times 10^{-4}$ Wb

cross-sectional area of field $A = 8\ \text{cm}^2$

Determine: The magnetic flux density in Tesla and in Gauss.

Answer 24

solution: $B = \dfrac{\Phi}{A}$

$= \dfrac{4 \times 10^{-4}}{8 \times 10^{-4}}\ \text{Wb/m}^2$

$= 0.5\ \text{T}$

$= 0.5 \times 10^4\ \text{G}$

$= 5000\ \text{G}$

The magnetic flux of a solenoid can be considerably increased by the insertion of a ferromagnetic core, without having to increase the number of turns or the current.

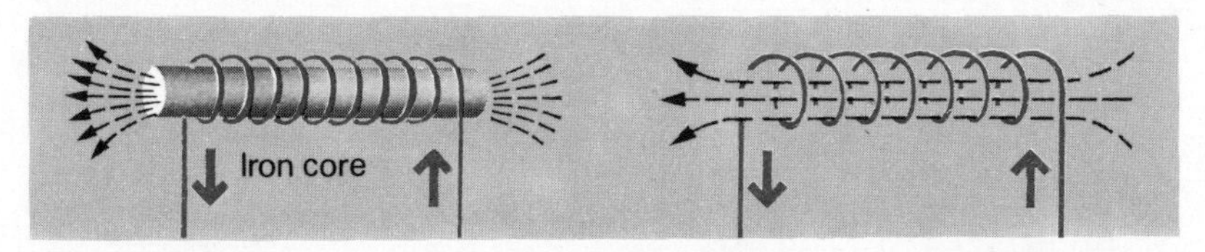

The reason for this increased magnetic flux is the aligning of the molecular magnet of the core material due to the magnetic field of the solenoid. The molecular magnets assist the solenoid magnetic field and by chosing a suitable core material it is possible to increase the magnetic flux many times over.

The factor by which the magnetic flux is increased is called the *relative permeability;* the symbol used for it is μ_r (pronounced mu-r). μ_r is a measure of how much the magnetic flux of the air-cored solenoid is increased by the insertion of the core material.

The numerical value of μ_r depends on the core material and its method of manufacture. For air or other non-magnetic materials $\mu_r = 1$. For magnetic materials, however, it can attain very high values. The values quoted in the table below are maxima since μ_r for ferromagnetic materials depends on the solenoid current.

Material	μ_r
Dynamo sheet steel	5000–8000
Mu metal (an alloy of Ni, Fe, Cu, Cr)	45 000
Alloy 1040 (of Ni, Fe, Cu, Mo)	100 000

QUESTION 25

A direct current flowing through a long air-cored solenoid generates a magnetic flux of 50×10^{-8} Wb.

With the current remaining constant, an iron core is inserted; this causes the magnetic flux to be increased to 4×10^{-3} Wb.

What is the relative permeability of the core material?

Answer 25

Solution: $\mu_r = \frac{\text{Magnetic flux with iron core}}{\text{Magnetic flux without iron core}}$

$$\mu_r = \frac{4 \times 10^{-3}}{50 \times 10^{-8}}$$

$$\mu_r = 8000$$

The magnetic properties of ferromagnetic materials are widely used in electrical engineering equipment.

Ferromagnetic materials are used for the magnetic flux carrying parts of electric machines. These parts can either be solid or built up from sheet laminations.

Transformer and transducer cores are built up from sheet material; for high-frequency applications cores are manufactured from compressed iron powder or from ferrites (see Lesson 1).

Final Test

Provided you have worked thoroughly through the lessons, you should have no difficulties in answering the questions below. Compare your answers with the solutions on page 64.

1 Which materials are attracted by a magnet?
(a) Iron, in the form of mild steel or cast iron
(b) Non-ferrous metals, e.g. nickel or cobalt
(c) Non-metallic materials

2 How do you locate the poles on a magnet?

3 What is the meaning of residual magnetism?

4 To set up magnetic fields do you
(a) only require ferromagnetic materials
(b) can they also be produced without these?

5 What are the smallest particles into which a magnet can theoretically be split?

6 If beads of ferromagnetic material are slipped over a straight current-carrying conductor will the magnetic field:
(a) be increased, because the magnetic flux is increased by ferromagnetic materials
(b) does it remain unchanged?

7 Is the term 'line of magnetic flux' used to indicate
(a) a movement
(b) a direction?

8 Using the 'left-hand rule' for a current-carrying conductor perpendicular to a magnetic field, does the outstretched thumb indicate:
(a) the direction of the force
(b) the direction of current flow
(c) the direction of the magnetic field?

9 Two identical freely movable solenoids are slipped on to a common core. The same current flows through both of them in

the same direction. Does the electromagnetic force acting between them cause them:
(a) to be attracted
(b) to be repelled?

10 You are given two iron bars which look externally absolutely alike. You have no other means at your disposal. How can you prove that both are permanent magnets?

11 Does the term 'ferromagnetic' mean that the material must contain iron?

12 When do two magnetic fields acting together cause the resulting magnetic field to be increased?

13 What is the meaning of relative permeability μ_r of a material?

14 What SI units are used for magnetic flux?

15 What is the effect of using a ferromagnetic core inside a current-carrying solenoid?

16 A current $I = 0.1$ A flowing through a solenoid generates a magnetic field with a magnetic flux density of 50×10^{-4} Tesla. The insertion of a ferromagnetic core raises the magnetic flux density to 1.2 Tesla. The current is now increased to three-times its former value causing the magnetic flux density generated by the air-cored solenoid to be increased to 150×10^{-4} Tesla, i.e. by 200%. The magnetic flux density of the solenoid containing the ferromagnetic core though rises only to 1.5 Tesla, i.e. by only 25%. How do you account for these different percentage increases of the magnetic flux density?

Solutions to Final Test Questions

1 All ferromagnetic materials are attracted by magnets. The materials mentioned under (a) and (b) are ferromagnetic; which also includes the ferrites although these are non-metallic.
(Lesson 1)

2 The poles will be found to be those regions of a magnet which exert the strongest attraction on ferromagnetic materials.
(Lesson 2)

3 Residual magnetism is the magnetism which a material retains after the source of magnetization has been removed. (Lesson 4)

4 Magnetic fields can also be produced without the use of ferromagnetic materials. (Lesson 8)

5 The smallest particles are the molecules, which we call molecular magnets. (Lesson 2)

6 (a) It increases in strength (its actual value depending on the relative permeability of the material of the beads).
(Lesson 25)

7 (b) It is used in a directional sense. (Lesson 13)

8 (a) The thumb indicates the direction of the force. (Lesson 17)

9 (a) With this arrangement poles of opposite polarity will face each other, i.e. the two coils will attract each other.
(Lesson 2, Lesson 22)

10 The two bars are brought close together. If in any one position repulsion is observed then both bars must be permanent magnets.

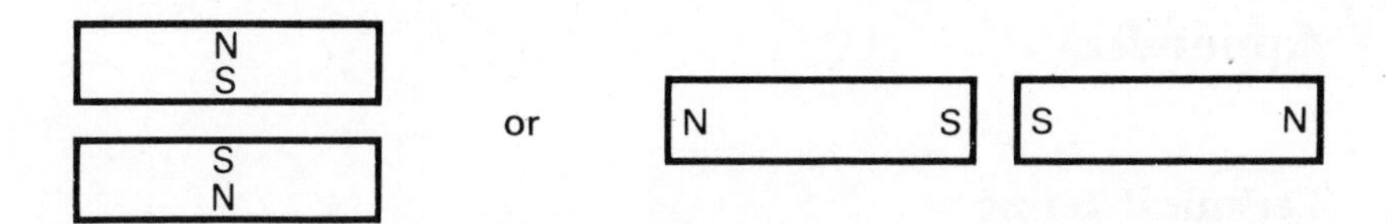

11 No, there are also iron-free ferromagnetic materials such as nickel or cobalt. (Lesson 1)

12 The resulting magnetic field of the two magnets is increased when the lines of magnetic flux of the two magnets act in the same direction. (Lesson 16)

13 μ_r indicates by how much the magnetic flux of an air-cored solenoid is increased by the insertion of a ferromagnetic core. (Lesson 25)

14 The SI unit Weber (or Volt-second) is used for magnetic flux. (Lesson 23)

15 The insertion of a ferromagnetic core will increase considerably the magnetic flux of a solenoid without having to raise the current or increase the number of turns. (Lesson 25)

16 The lower increase of only 25% in the magnetic flux density in the iron-cored solenoid is due to magnetic saturation of the core material. This makes it impossible to increase the magnetic flux density any further. (Lesson 4)

Appendix

Technical terms

alloy	substance obtained by mixing several metals by melting (Lesson 1)
direct current	an electric current always flowing in the same direction (Lesson 8)
dynamic	moving (Lesson 16)
electron	smallest fundamental particle (Lesson 11)
ferrite	derived from ferrum, Latin for iron (Lesson 1)
ferromagnetic	magnetic similar to iron
Gauss	Karl Friederick Gauss (1777–1855) German Physicist. Unit of magnetic flux density (Lesson 24)
homogeneous	uniform or of the same kind (Lesson 21)
insulating material	materials in which electric conduction is either completely absent or very low (Lesson 9)
Maxwell	James C. Maxwell (1831–1879) British Physicist. Unit of magnetic flux (Lesson 23)

oxide — the chemical compound of an element with oxygen (Lesson 1)

permanent — from the Latin permanere—to persist (Lesson 1)

permeability — a magnetic parameter giving the ratio between magnetic flux density and magnetic field strength (Lesson 25)

remanence — residual magnetism; from the Latin remanere—to remain (Lesson 4)

SI — abbreviation for 'Système International d'Unités' = International system of units (Lesson 23)

sinter — combining materials by heating them until they become doughy just below their melting point (Lesson 1)

stabilizing — bringing into equilibrium (Lesson 12)

Tesla — Nicola Tesla (1856–1943); Croat Physicist. Unit of magnetic flux density (Lesson 24)

transformer — converter of electrical energy from one level to another (Lesson 4)

Weber — Wilhelm Weber (1804–1891) German Physicist. Unit of magnetic flux (Lesson 23)

Maxwell's eqns.
E. G. Thomas
& A. J. Meadows